Animals on the Farm

# Ducks

by Wendy Strobel Dieker

Bullfrog
Books

# Ideas for Parents and Teachers

Bullfrog Books give children practice reading informational texts at the earliest levels. Repetition, familiar words, and photos support early readers.

## Before Reading

• Discuss the cover photo with the class. What does it tell them?

• Look at the picture glossary together. Read and discuss the words.

## Read the Book

• "Walk" through the book and look at the photos. Let the child ask questions.

• Read the book to the child, or have him or her read independently.

## After Reading

• Prompt the child to think more. Ask: Would you like to raise ducks? Would a duck make a good pet? Why or why not?

Bullfrog Books are published by Jump!
5357 Penn Avenue South
Minneapolis, MN 55419
www.jumplibrary.com

Library of Congress Cataloging-in-Publication Data
Dieker, Wendy Strobel.
Ducks / by Wendy Strobel Dieker.
    p. cm. — (Bullfrog books: animals on the farm)
Includes index.
Summary: "Ducks narrate this photo-illustrated book describing the body parts and behavior of ducks on a farm. Includes picture glossary" —Provided by publisher.
ISBN 978-1-62031-002-1 (hardcover : alk. paper)
ISBN 978-1-62031-629-0 (paperback)
1. Ducks—Behavior—Juvenile literature. 2. Ducks as pets—Juvenile literature. I. Title.
SF505.3.D54 2013
636.5'97--dc23
                                    2012008223

Thanks, Joe Showalter! - W.D.

Series Editor: Rebecca Glaser
Series Designer: Ellen Huber
Production: Chelsey Luther

Photo Credits: Dreamstime.com, 1, 4, 5, 16–17, 23bl, 24; Getty, 7, 13, 14–15; iStockphoto, 20 (all); Shutterstock, 3b, 3t, 8, 12, 14, 21, 22, 23br; SuperStock, 6, 8–9, 10–11, 18–19, 23tr 23tl

Printed in the United States of America at Corporate Graphics in North Mankato, Minnesota

# Table of Contents

# Ducks on the Farm

I am a duck.
I live on a farm.

5

Do you see my webbed feet?

They are like paddles.

They help me swim.

feathers

Do you see my oily feathers?
The oil helps me float.

drake

Do you see his
colorful feathers?

He is called
a drake.

Females are
just called ducks.

Do you see my bill?
I scoop up bugs
in the pond.

bill

12

I even eat frogs!

Do you hear the ducks quack?

Mama ducks are loud.

Drakes make a quiet sound.

Do you see
my nest?

I lay an egg
most days.

Do you see my fuzzy ducklings?

**They cannot swim yet.**

The ducklings are two weeks old.
I teach them to swim.
Quack! Quack! Stay with mama!

# Parts of a Duck

**feathers**
Light, fluffy parts that cover a bird's body.

**bill**
A duck's beak; the nose and mouth are part of the bill.

**webbed feet**
Feet that have toes connected by skin; webbed feet help ducks swim.

# Picture Glossary

**drake**
The name for a male duck.

**duckling**
The name for a baby duck.

**duck**
The name for a female duck.

**oily**
Covered in a liquid that does not mix with water; ducks have oily feathers.

# Index

# To Learn More

Learning more is as easy as 1, 2, 3.

1) Go to www.factsurfer.com

2) Enter "duck" into the search box.

3) Click the "Surf" button to see a list of websites.

With factsurfer.com, finding more information is just a click away.